HUMAN COMPUTER

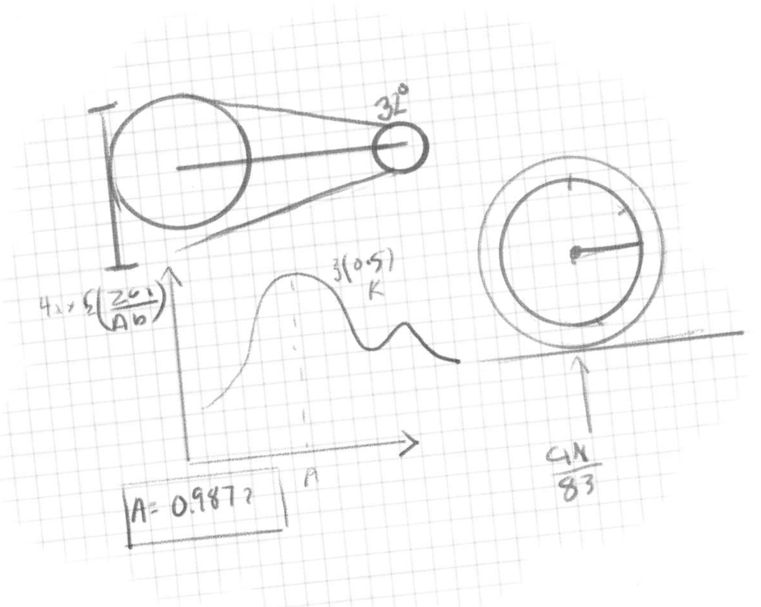

Mary Jackson

ENGINEER

ANDI DIEHN

Illustrated by Katie Mazeika

AS A GIRL, MARY WANTED TO BE AN ENGINEER.
BUT BECAUSE SHE WAS BLACK, SOME SAID, "NOT YOUR CAREER."

SEGREGATION TOOK AWAY HER PATH OF CHOICE,
UNTIL SHE FOUGHT BACK AND RAISED HER VOICE.

DEMANDING TO BE LET INTO THE CLASSES SHE NEEDED,
AND THROUGH PERSISTENCE AND HARD WORK, SHE SUCCEEDED!

SHE BECAME NASA'S FIRST BLACK, FEMALE ENGINEER!
THEN, SHE MADE SURE THE PATH FOR OTHERS WAS CLEAR.

Even as a young girl, Mary Jackson loved math and science. **She loved learning how things worked.** And she loved helping people.

Mary wanted to be
an engineer and to
solve problems with
math and science.

But Mary grew up in a time when black people and white people were kept apart by something called segregation. Black children weren't always taught the same subjects as white children.

There were very few black engineers. And even fewer female engineers.

Mary still loved learning math and science. **She wanted to know how everything worked.**

Mary beat the odds and went to college! She became a math teacher and shared her love of numbers with her black students.

Mary shared her learning with other little girls, too. She was a Girl Scout leader!

She loved to build wind tunnels and test model airplanes with her scouts.

A few years later, after Mary had a baby, she went to work as a clerk. She answered the phone and wrote messages. **She was good at her job, but she missed learning how different things worked.**

R=
R

So, Mary got a job at the Langley Memorial Aeronautical Laboratory as a human computer.

Engineers at Langley were working to improve airplanes. They needed human computers who were very good at math to help them.

Mary was happy to work with numbers again.

Mary loved her job, but there was a big problem.

Langley was segregated.

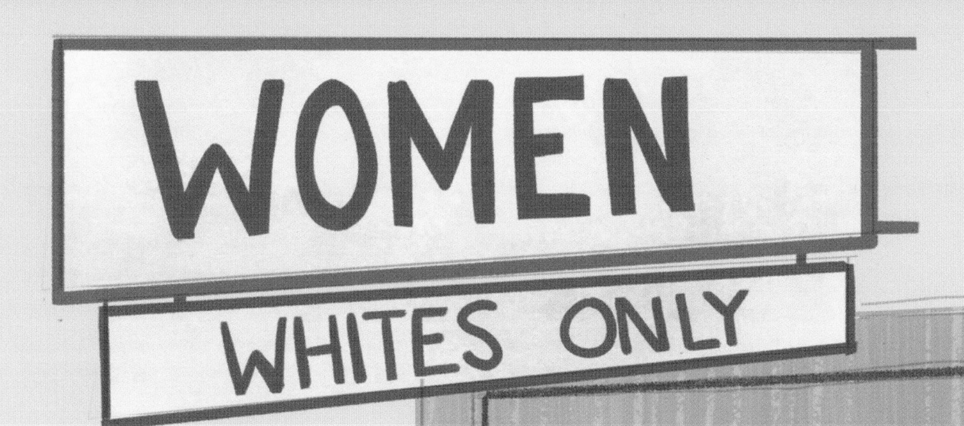

Mary grew more and more angry. She was tired of being treated differently.

There were many people who thought segregation was wrong. An engineer named Kazimierz Czarnecki asked Mary to come work with him. Kaz thought everyone should all work together.

And Mary agreed! She went to work as a computer for engineers who were studying things that traveled at supersonic speed, or faster than the speed of sound.

K.CZARNECKI
M.JACKSON

Mary wanted to be an engineer more than ever. All she needed to do was take classes at a local school. But there was one problem.

The school was segregated. Black people weren't allowed to go to school there.

Mary had to ask the city to give her special permission to take classes at the white school.

Finally, the city said yes.

Mary studied hard and became NASA's first African American female engineer!

M.JACKSON

Mary worked as an engineer for many years. She helped design the spacecraft that first carried men to the moon!

Mary had achieved her dream. Now, she wanted to help other people achieve their dreams. After all, no one had expected Mary to become the first African American female engineer at NASA!

Mary Jackson
Manager
OFFICE of EQUAL
OPPORTUNITY PROGRAMS

Fig.5

And she knew she wouldn't be the last.

Spaghetti Strength!

Engineers work hard to make sure the structures they design can support the weight they're supposed to support! You don't need to build a spaceship to test some structural strength.

What You Need: a box of spaghetti, marshmallows

First, test the strength of your spaghetti pieces by trying to break one, then a bunch of 10, then 20.

Which bundle breaks most easily? Which is the strongest bundle?

Make a tower by attaching strands of spaghetti with the marshmallows.

How high can you make a tower go? Can you create a structure that will hold up a hardcover book?

Make more towers!

Try to build a tower that is both tall and strong. How many strands of spaghetti do you need to use? How wide does the base of the tower need to be? What can you add to make your tower stronger? These are all questions engineers ask when they are designing things!

QUOTE CONNECTIONS!

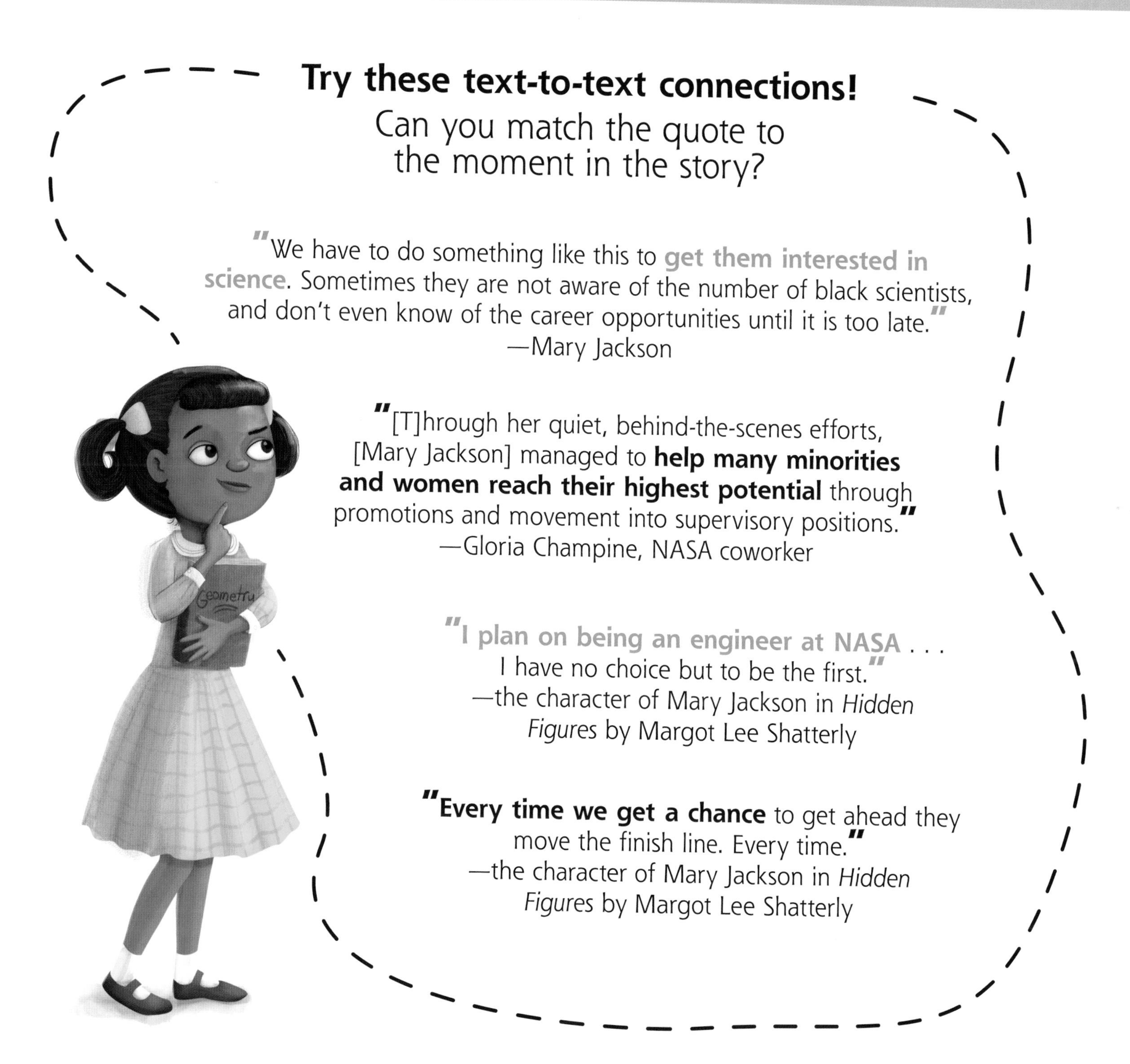

Try these text-to-text connections!

Can you match the quote to the moment in the story?

"We have to do something like this to **get them interested in science**. Sometimes they are not aware of the number of black scientists, and don't even know of the career opportunities until it is too late."
—Mary Jackson

"[T]hrough her quiet, behind-the-scenes efforts, [Mary Jackson] managed to **help many minorities and women reach their highest potential** through promotions and movement into supervisory positions."
—Gloria Champine, NASA coworker

"**I plan on being an engineer at NASA** . . . I have no choice but to be the first."
—the character of Mary Jackson in *Hidden Figures* by Margot Lee Shatterly

"**Every time we get a chance** to get ahead they move the finish line. Every time."
—the character of Mary Jackson in *Hidden Figures* by Margot Lee Shatterly

TIMELINE

1921 Mary Winston is born in Hampton, Virginia. She later marries Levi Jackson.

1942 Mary earns a bachelor's degree in mathematics and physical science from Hampton Institute.

1952 Mary goes to work as a computer at the National Advisory Committee for Aeronautics (NACA).

A high-speed wind tunnel at Langley, from the outside, 1936

1953 Mary takes a job working for engineer Kazimierz Czarnecki.

1956 Mary petitions the city of Hampton to allow her to attend engineering classes at the segregated Hampton High School.

TIMELINE

1958 Langley becomes part of NASA. Mary becomes the first female black engineer at NASA.

Mary Jackson in 1977

1979 Mary leaves the engineering department at NASA to become an administrator and equal opportunity specialist.

1985 Mary retires from NASA.

Langley Research Center, circa 2011

Mary Jackson in 1977

GLOSSARY

clerk: someone who works in an office, keeping records.

engineer: a person who uses science, math, and creativity to design and build things.

computer: a device for storing and working with information. Before digital computers, people who worked with numbers were often called human computers.

NASA: National Aeronautics and Space Administration, the U.S. organization in charge of space exploration.

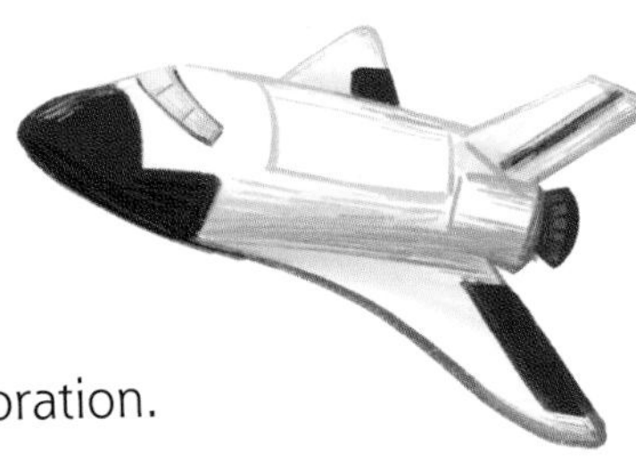

permission: approval for something to happen.

scout: a member of a group that promotes character, outdoor activities, good citizenship, and service to others.

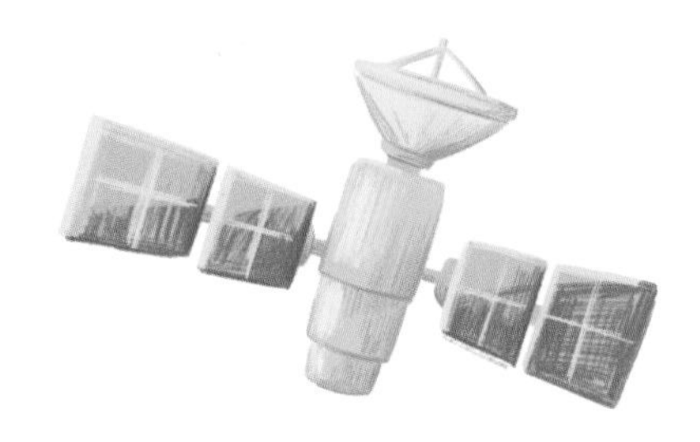

segregation: the policy of keeping people of different races separate from each other.

spacecraft: a spaceship that explores the solar system and sends data back to Earth.

supersonic speed: a speed that's faster than the speed of sound.

wind tunnel: a tunnel built so airplanes can be safely tested in high winds.